Practice Exam for the Civil PE Examination

Breadth + Transportation Depth

Indranil Goswami, Ph.D., P.E.

September 2015

Second printing September 2015

Preface

In January 2015, the official (NCEES) syllabus for the PE-Civil examination underwent a significant realignment. There was a significant departure from the previous structure which placed approximately equal emphasis on the five areas of practice. In the breadth exam, the 40 questions were approximately equally distributed among Construction, Geotechnical, Structural, Transportation and Water Resources. In the current syllabus for the Breadth (AM) exam, Transportation has been significantly deemphasized while there seems to be more emphasis on Construction.

The new depth (PM) syllabi have also gone through reorganization as well as addition of specific subtopics under various categories.

These practice exams were developed *after* the syllabus went through the aforementioned reorganization and are therefore consistent with the same.

This full-length practice exam contains 40 breadth (AM) questions + 40 depth (PM) questions in the area of TRANSPORTATION ENGINEERING. It should be taken under as near exam conditions as possible, preferably at the point when you think your exam review is complete and you are ready to take a simulated test to assess the level of your preparation. You should even go so far as to ask someone else to detach the questions from the solutions, so that you don't have any temptation to peek.

All the best for the upcoming PE exam,

Indranil Goswami

P.S. In this second printing, errors discovered to date have been corrected.

Table of Contents

BREADTH EXAM QUESTIONS 001 - 040	05-27
TRANSPORTATION DEPTH QUESTIONS 401 - 440	29-44
BREADTH EXAM ANSWER KEY	48
BREADTH EXAM SOLUTIONS 001 - 040	47-59
TRANSPORTATION DEPTH ANSWER KEY	62
TRANSPORTATION DEPTH SOLUTIONS 401 - 440	61-73

BREADTH EXAM
FOR THE
CIVIL PE EXAM

The following set of 40 questions (numbered 001 to 040) is representative of a 4-hour breadth (AM) exam according to the syllabus and guidelines for the Principles & Practice (P&P) of Civil Engineering Examination (updated January 2015) administered by the National Council of Examiners for Engineering and Surveying (NCEES). The exam is weighted according to the official NCEES syllabus (2015) in the following subject areas – Construction, Geotechnical, Structural, Transportation and Water & Environmental. Copyright and other intellectual property laws protect these materials. Reproduction or retransmission of the materials, in whole or in part, in any manner, without the prior written consent of the copyright holder, is a violation of copyright law.

The time allocated for this set of questions is 4 hours.

001

The tables below show historical data on traffic counts for a bridge, averaged by day of the week and month. If a daily count, conducted on a Wednesday in April is 19,545, the AAWT (Average Annual Weekday Traffic) for planning purposes is most nearly

    A. 16,300
    B. 17,250
    C. 22,830
    D. 23,440

Day of the week	ADT
Sunday	12,760
Monday	18,985
Tuesday	20,765
Wednesday	19,882
Thursday	20,349
Friday	16,889
Saturday	13,725
TOTAL	123,355

Month	ADT
January	17,756
February	16,772
March	19,674
April	21,983
May	20,935
June	16,783
July	15,887
August	16,785
September	19,836
October	19,356
November	20,128
December	19,785
TOTAL	225,680

002

The boundaries of a site form a triangle as shown below. The area of the site (acres) is most nearly

    A. 16
    B. 24
    C. 32
    D. 48

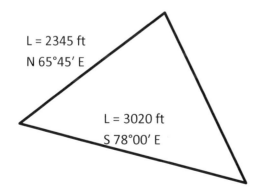

CIVIL PE SAMPLE EXAMINATION: TRANSPORTATION © Copyright Indranil Goswami 2015

003

A reinforced concrete pipe of 36 inch outer diameter connects two manholes MH-1 and MH-2 as shown in the figure below. At station 13 + 05.10, the ground surface has a low point elevation of 242.35 ft. At this location, the soil cover (feet) is most nearly:

    A. 3.95
    B. 4.12
    C. 4.56
    D. 4.72

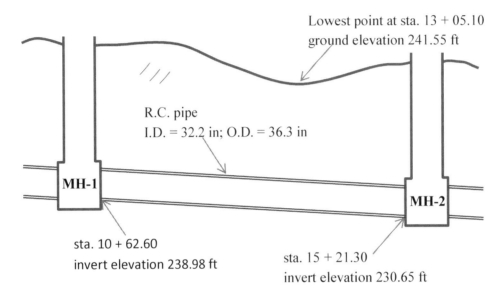

004

A roadside gutter is in the shape of a symmetric v-channel with 3H:1V sideslopes. The gutter is to be lined with a 3 inch thick concrete liner as shown. If the concrete material + placement cost is $232/yd^3, then the cost of constructing the gutters ($/mile) is most nearly

    A. 36,000
    B. 41,000
    C. 72,000
    D. 82,000

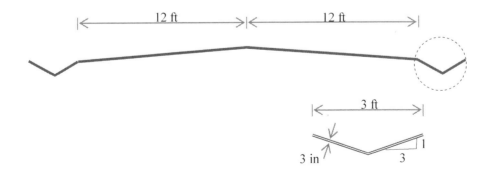

005

The activity on node network for a project is as shown below. All relationships are finish to start unless otherwise indicated. The table on the right shows pertinent data. The early start date (weeks) for activity F is:

    A. 15
    B. 16
    C. 17
    D. 18

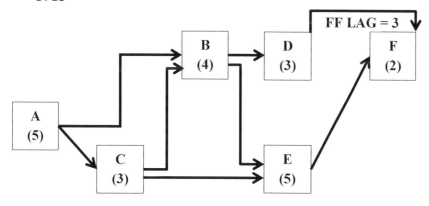

006

A parabolic vertical curve must connect a tangent of slope +5% to another of slope -3% as shown below. The two tangents intersect at a point at station 11 + 45.20 and at elevation 310.56 ft. A sewer (circular CIP) with crown elevation 302.65 ft exists at station 12 + 30.05. If minimum soil cover of 30 inches is required above the sewer pipe, the required length of curve (feet) is most nearly

    A. 510
    B. 570
    C. 635
    D. 815

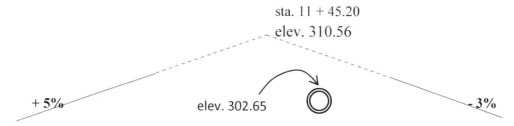

007

A cantilever retaining wall is supported by a 3-foot thick footing as shown. The drains behind the wall become clogged and groundwater rises to the top of the horizontal backfill. The total horizontal earth pressure resultant (lb/ft) acting on the retaining wall is most nearly:
A. 22,300
B. 20,650
C. 18,770
D. 13,300

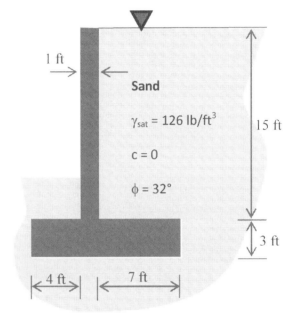

008

A 10-ft layer of varved clay is overlain by a 12 ft thick sand layer, as shown. The water table is originally at a depth of 5 ft below the ground surface. Prior to construction, the water table is lowered by 7 ft, to the bottom of the sand layer. Three months after lowering the water table, settlement (inches) due to consolidation of the clay layer is most nearly
A. 0.9
B. 1.9
C. 2.6
D. 4.0

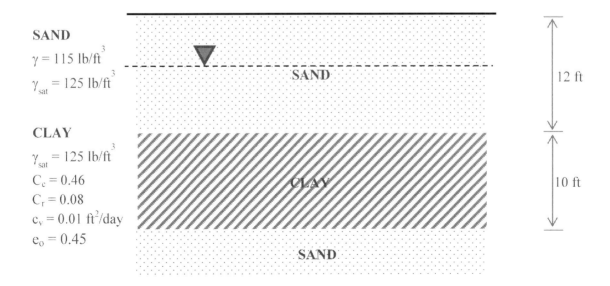

009

Identify the correct shape of the bending moment diagram for the beam loaded as shown below.

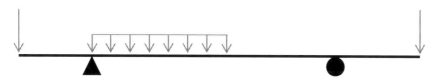

A.

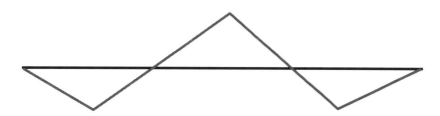

B.

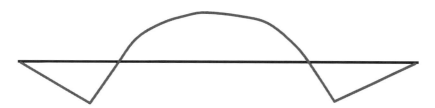

C.

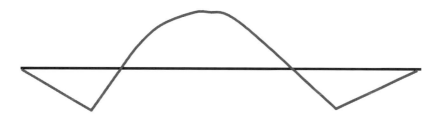

D.

010

A rectangular post is subject to an eccentric load P as shown. The maximum compressive stress (MPa) is most nearly

    A. 1.0
    B. 1.6
    C. 2.6
    D. 4.0

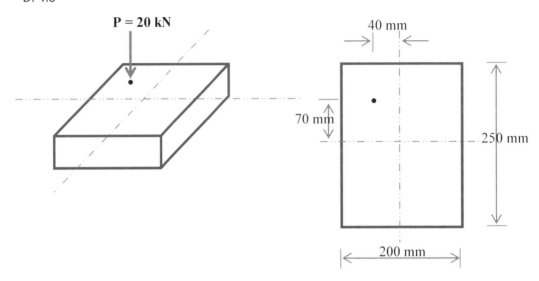

011

A trapezoidal open channel conveys flow at a uniform depth of 5 ft as shown below. The Manning's n = 0.015. The bottom width of the channel = 20 ft and longitudinal slope of the channel floor is 0.8%.

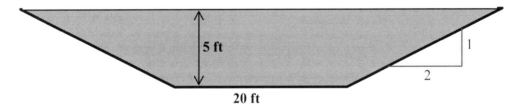

The flow rate (MGD) conveyed by the channel is most nearly:

    A. 500
    B. 1000
    C. 2000
    D. 3000

012

A 5 ft x 5 ft square footing transfers a column load of 140 kips to a sandy soil as shown. The depth of the footing is 3 ft. The factor of safety against general bearing capacity failure is most nearly

  A. 3.3
  B. 2.7
  C. 2.0
  D. 1.3

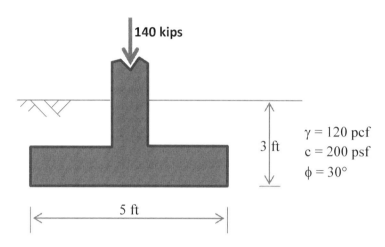

$\gamma = 120$ pcf
$c = 200$ psf
$\phi = 30°$

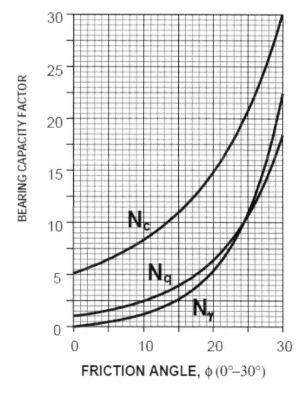
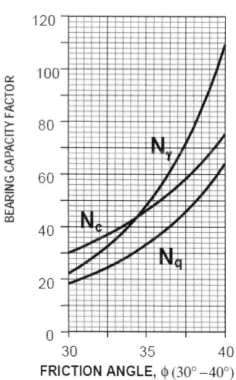

013

A wall panel is composed of plywood sheathing supported by 2x6 studs (nominal dimensions 1.5 in x 5.5 in) spaced every 28 inches as shown. The studs are supported by longitudinal members every 10 ft. The wall experiences a normal wind pressure of 30 psf. The maximum bending stress (lb/in^2) in the studs is most nearly:

A. 800
B. 1000
C. 1200
D. 1400

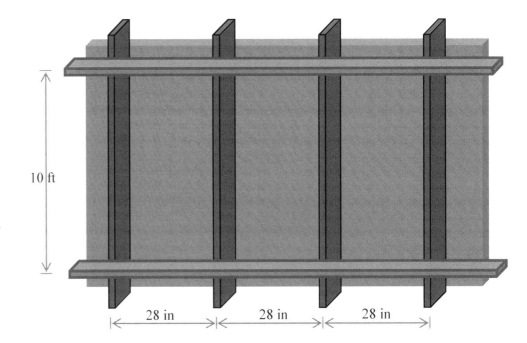

014

The table below shows cross section areas of cut and fill recorded at 5 stations spaced at 100 ft.

Station	Area (ft^2)	
	CUT	FILL
0 + 0.00	245.0	123.5
1 + 0.00	312.5	76.3
2 + 0.00	411.5	0.0
3 + 0.00	234.5	88.4
4 + 0.00	546.2	214.5

The net earthwork volume (yd^3) between stations 0 + 0.00 and 4 + 0.00 is most nearly

A. 3640 (cut)
B. 3780 (cut)
C. 3640 (fill)
D. 3780 (fill)

015

A site needs soil compacted to 90% of the Proctor maximum dry density. The results of the Proctor test are shown below.

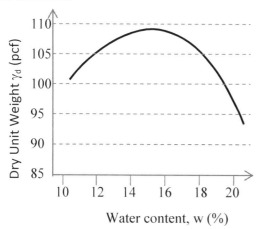

The volume of the embankment is 50,000 ft³. If borrow soil is available at γ = 120 pcf and moisture content = 14%, the volume of soil (yd³) needed from the borrow pit is most nearly:

  A. 1,400
  B. 1,730
  C. 2,080
  D. 2,800

016

Isohyets showing precipitation depth are shown in the figure. The accompanying table shows total area enclosed by each close contour. The average precipitation depth (inches) is most nearly:

  A. 0.46
  B. 0.51
  C. 0.58
  D. 0.62

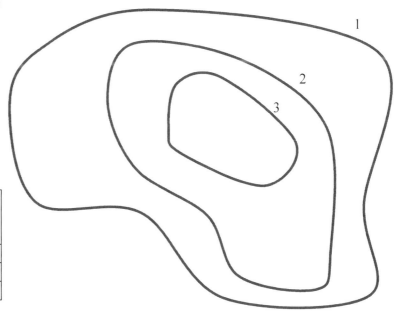

Contour	Area enclosed (acres)	Precip. depth (inches)
1	268	0.2
2	121	0.6
3	45	0.9

017

A 24 inch diameter reinforced concrete pipe (C = 100) conveys water a flow rate = 12.5 ft³/sec. The head loss due to friction (feet per mile) is most nearly:
  A. 11.5
  B. 14.6
  C. 16.3
  D. 18.3

018

Runoff flow from a development is held in a detention pond until it is 78% full, at which point, it empties through a weir. The capacity of the pond is 760,000 gallons. If after a rainfall event, the average rate of inflow into the pond occurs at 2 ft³/sec, the length of time (hours) before the pond starts to empty is most nearly:
  A. 7.5
  B. 9.3
  C. 11.0
  D. 13.7

019

For the truss shown below, the axial force (kips) in member CD is most nearly:
  A. 125 (tension)
  B. 125 (compression)
  C. 150 (tension)
  D. 150 (compression)

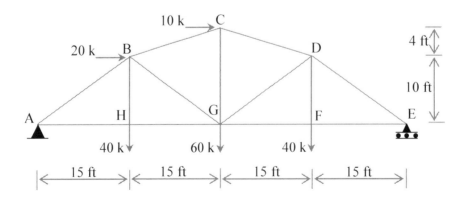

020

A 6-inch thick riprap layer is used as protection for the earth slope (θ = 30°) shown below. The factor of safety for slope stability is most nearly:

    A. 1.25
    B. 1.35
    C. 1.45
    D. 1.55

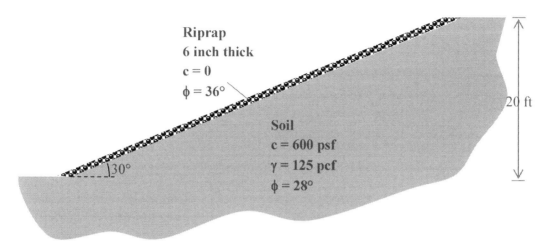

021

Particle size distribution of a soil sample is summarized in the table below. The fines were tested using Atterberg apparatus to obtain the following results:

Liquid limit   45
Plastic limit   21

Sieve size	% finer
1.0 inch (25.4 mm)	100
0.5 inch (12.7 mm)	92
No. 4 (4.75 mm)	75
No. 10 (2.00 mm)	62
No. 40 (0.425 mm)	52
No. 100 (0.15 mm)	45
No. 200 (0.075 mm)	28

The classification of the soil according to the USCS is:

    A. GW
    B. SM
    C. SC
    D. GC

022

A soil sample yields the following results:

Mass of wet soil = 1685 g

Volume of wet soil = 855 cc

Mass of soil after oven drying = 1418 g

If the specific gravity of soil solids is taken as 2.65, the void ratio is most nearly:

    A. 0.3
    B. 0.4
    C. 0.5
    D. 0.6

023

2x6 stud columns (actual dimensions 1.5 in x 5.5 in) are connected to plywood sheathing as shown. The modulus of elasticity of timber is $E = 1.5 \times 10^6$ lb/in^2. The Euler buckling load (kips) for each column is most nearly:

    A. 9
    B. 15
    C. 21
    D. 26

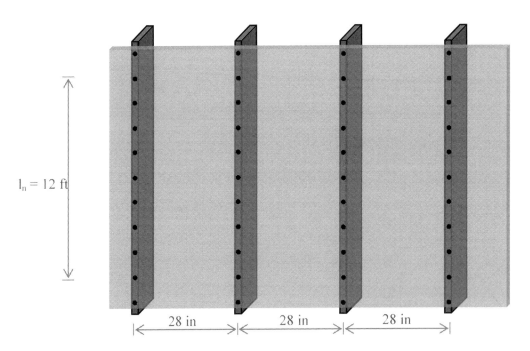

024

Water stored in a large reservoir (surface elevation 324.5 ft above sea level) empties through a 24 inch diameter pipe as shown. The far end of the pipe is at elevation 295.8 ft above sea level. The discharge (ft³/sec) through the pipe is most nearly:

    A. 50
    B. 38
    C. 32
    D. 22

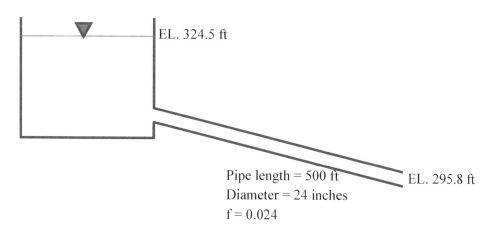

Pipe length = 500 ft
Diameter = 24 inches
f = 0.024

EL. 324.5 ft
EL. 295.8 ft

025

Surface runoff from a catchment area has the longest hydraulic path as shown in the figure below. Time for sheet flow, t_s = 5 minutes and for ditch flow t_d = 13 minutes. A collector pipe, in which flow occurs at an average velocity of 5 feet/sec, of length 1200 ft has an inlet at point A and discharges into a main sewer at point B. Intensity-duration-frequency curves are obtained from historical precipitation data. The design intensity (in/hr) to be used for the design of the sewer mains for a 20 year storm is most nearly:

    A. 1.5
    B. 1.8
    C. 2.1
    D. 2.4

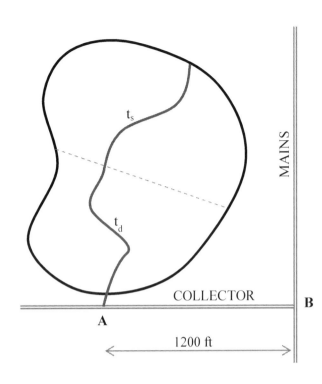

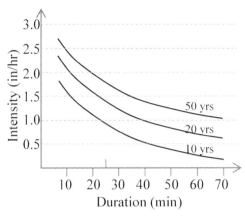

026

A circular horizontal curve has the coordinates (feet) of the PC, PI and PT as shown on the figure.

The degree of curve (degrees) is most nearly:
  A. 10
  B. 13
  C. 16
  D. 19

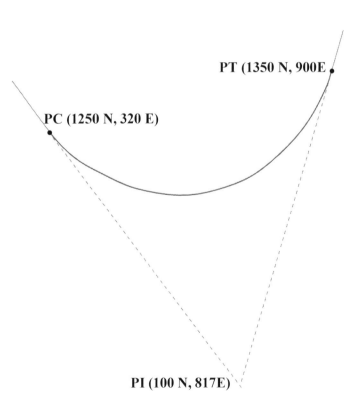

027

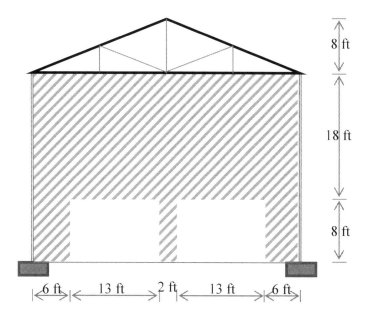

A 110 ft long warehouse shed has metal sheathing forming the sides (no openings) and front and back (with two door openings on each surface) and the roof. The total surface area of metal sheathing (sq. ft.) is most nearly:

    A. 12,652
    B. 12,456
    C. 12,331
    D. 12,123

028

A crane with a 40 ft boom is used to lift a 4 ton load as shown. The total weight of the crane and ballast is 4.5 tons acting at the effective location indicated as CG on the figure. The weight of the boom is 800 lb. Each outrigger leg is supported by a circular pad with a diameter = 3 feet

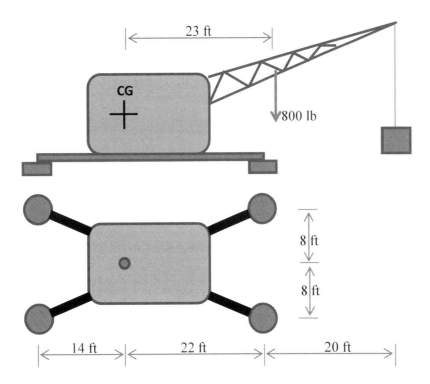

The maximum soil pressure (lb/ft^2) under the outrigger pads is most nearly:
- A. 1,200
- B. 1,600
- C. 2,000
- D. 2,400

029

Which of the following statements is true about deflection of concrete beams?
- I. The deflection is calculated using the moment of inertia of the uncracked section
- II. The deflection is calculated using a moment of inertia equal to half the gross moment of inertia
- III. The deflection is calculated using the moment of inertia of the cracked section
- IV. The deflection is calculated using a moment of inertia less than the gross moment of inertia

- A. I and IV
- B. IV only
- C. II only
- D. II and III

030

The figure shows a stress-strain diagram based on a tension test of a steel test coupon.

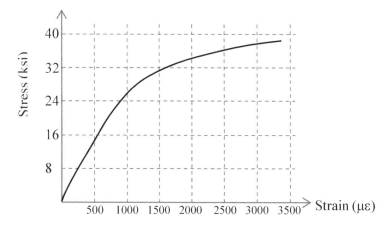

The yield stress (kip/in^2) is most nearly

   A. 26
   B. 30
   C. 34
   D. 38

031

A layer of coarse sand (thickness = 15 feet) supports a mat foundation that exerts a net uniform pressure of 600 psf at a depth of 3 feet below the surface as shown. The compression of the sand layer (inches) is most nearly:

   A. 0.1
   B. 0.2
   C. 0.3
   D. 0.4

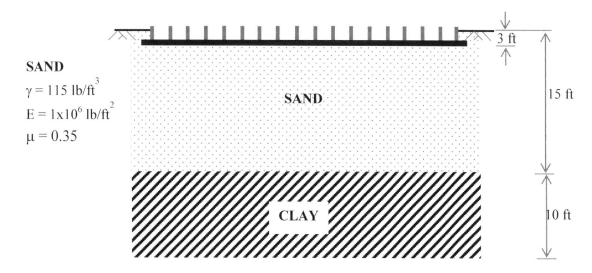

032

Which of the following statements about total float (TF) in a CPM network is not true? Use the following nomenclature – ES (early start), EF (late finish), LS (late start), LF (late finish), D (duration).

I. TF = LF – ES – D
II. TF = LS – EF – D
III. TF = LS – EF + D
IV. TF = LS – ES

A. I
B. II
C. III
D. IV

033

Which of the following statements is/are true?

I. For long term stability analysis of clay slopes, results of the CD triaxial test must be used.
II. The CD triaxial test takes longer to perform than the UU test.
III. The UU triaxial test takes longer to perform than the CD test
IV. Pore pressure measurements must be made during the CD test

A. I, II only
B. II, III only
C. III only
D. all of them

034

A test strip shows that a steel-wheeler roller, operating at 3 mph, can compact a 0.5 ft. layer of material to a proper density in four passes. The width of the drum is 8.0 ft. The roller operates 50 min per hour. The number of rollers required to keep up with a material delivery rate of 540 bank cubic yards/hr is most nearly: (1 bank cubic yard = 0.83 compacted cubic yard):

A. 4
B. 3
C. 2
D. 1

035

A simply supported steel beam carries a single concentrated load at midspan as shown. The beam as the following properties: Area A = 29.4 in^2; I$_x$ = 1490 in^4; I$_y$ = 186 in^4; Z$_x$ = 198 in^3; Z$_y$ = 54.9 in^3; S$_x$ = 175 in^3; S$_y$ 35.7 = in^3.

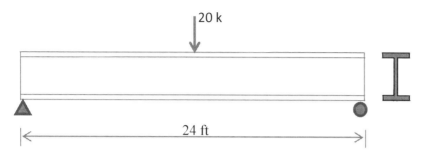

The maximum deflection (inches) is most nearly:
- A. 0.25
- B. 0.50
- C. 0.75
- D. 0.90

036

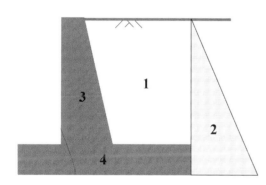

	Component	Resultant Force
1	Backfill soil	12,600 lb/ft
2	Active earth pressure	5,400 lb/ft
3	Concrete wall stem	3,500 lb/ft
4	Concrete wall footing	4,500 lb/ft

A concrete retaining wall has a level backfill behind it. The coefficient of friction between the wall footing and the soil is 0.6. The table to the right summarizes the forces acting on the retaining wall. The factor of safety against sliding is most nearly:
- A. 0.9
- B. 1.6
- C. 2.3
- D. 2.9

037

The 'first flush' runoff depth (assumed to be 1 inch) from a 120 acre watershed collects in a 2 acre detention pond. The sediment load carried by the runoff is 5 g/L. Bulk density of sediment is 80 lb/ft^3. The loss of depth (inches) per rainfall event is most nearly:

    A. 0.125

    B. 0.25

    C. 0.50

    D. 0.75

038

Several cylindrical steel samples are tested in a Universal Testing Machine to obtain the results obtained below:

Sample	Diameter (in)	Sample length (in)	Breaking load (lb)	Elongation at Failure (in)
1	0.504	5.66	9050	0.412
2	0.498	7.34	8865	0.516
3	0.509	7.55	9235	0.543
4	0.503	8.12	9110	0.581
5	0.512	7.12	9565	0.505

The average breaking strain ($\mu\varepsilon$) is most nearly:

    A. 38,500

    B. 51,200

    C. 71,500

    D. 78,300

039

A circular conduit of diameter 48 inches conveys water at a depth of 30 inches as shown below. The interior of the concrete pipe is coated to yield n = 0.014 (assumed constant with varying depth).

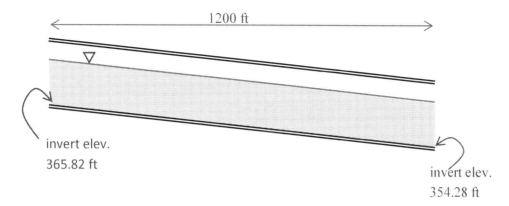

The velocity (ft/sec) is most nearly:
    A. 8.2
    B. 9.3
    C. 10.4
    D. 11.3

040

Which of the following techniques are commonly used for construction adjacent to historic structures?
    I. Underpinning
    II. Anchor rod and deadman
    III. Slurry walls
    IV. Compaction piles

A. I and II
B. I and III
C. II and III
D. II and IV

THIS IS THE END OF THE BREADTH EXAM

TRANSPORTATION DEPTH EXAM
FOR THE
CIVIL PE EXAM

The following set of 40 questions (numbered 401 to 440) is representative of a 4-hour depth (PM) exam for TRANSPORTATION according to the syllabus and guidelines for the Principles & Practice (P&P) of Civil Engineering Examination (updated January 2015) administered by the National Council of Examiners for Engineering and Surveying (NCEES). Copyright and other intellectual property laws protect these materials. Reproduction or retransmission of the materials, in whole or in part, in any manner, without the prior written consent of the copyright holder, is a violation of copyright law.

The time allocated for this set of questions is 4 hours.

401

A freeway segment has the following data:
- Number of lanes = 3
- Design hourly volume = 4700 vph
- 5% trucks
- 3% buses
- Lane width = 11 ft
- Shoulder width = 6 ft
- Average grade = +3%
- Commuter traffic; PHF = 0.92
- 4 full cloverleaf interchanges in the 6 mile stretch adjacent to and centered on the segment.

The level of service is:

A. A
B. B
C. C
D. D

402

A stretch of highway experienced 12 crashes during 2008. ADT during 2008 was 15,600. Due to the significant number of crashes, the following three strategies were put in place:

1. Lane widening – expected CMF = 0.90
2. Parking restrictions – expected CMF = 0.75
3. Lighting improvement – expected CMF = 0.80

If traffic grows by approximately 3% every year, the number of crashes expected in 2010 is most nearly:

A. 3
B. 5
C. 7
D. 9

403

The findings of a speed survey are summarized below.

Speed Interval (mph)	Frequency
20 – 25	2
25 – 30	9
30 – 35	16
35 – 40	28
40 – 45	19
45 – 50	11
50 – 55	3

The 85th percentile speed, (miles per hour) is most nearly
    A. 43.2
    B. 44.7
    C. 46.1
    D. 45.5

404

The table below summarizes accident data for 4 different highway segments. For all segments, accident data has been summarized from 3 years. If the budget has only enough money to fund improvements on two segments, which two segments qualify for funding?

Segment	Length (miles)	No. of accidents	ADT
1	0.8	13	13,000
2	0.4	10	9,500
3	0.3	5	15,000
4	0.5	7	9,000

    A. Segments 2 and 4
    B. Segments 2 and 3
    C. Segments 1 and 3
    D. Segments 1 and 4

405

A car accelerates uniformly from rest to its peak speed of 70 mph. The acceleration rate is 8 mph/sec. After cruising a certain distance at peak speed, the vehicle brakes to rest, decelerating uniformly at 10 mph/sec. If the total distance traveled is 0.5 miles, the average running speed (mph) is most nearly:

    A. 48.7
    B. 51.2
    C. 53.6
    D. 57.4

406

A trip generation model is created from empirical data. The primary variables are X (households per auto), Y (persons per household) and Z (household income in thousands of dollars). The number of daily trips per household (T) is given by

$$T = 2.3X + 0.6Y + 0.02Z$$

The total number of daily trips from a community of 250 households with 3.7 persons and 1.7 automobiles per households and an average household income of $134,000 is most nearly:

    A. 1050
    B. 1190
    C. 1310
    D. 1560

407

A signalized intersection has a 96 second cycle. The signal cycle has 4 phases, results for which are listed in the table below. Amber clearance time between successive phases = 4 seconds. Assume lost time per cycle = 3 seconds.

Phase	Description	Green Time (sec)	Critical Lane Group		
			No. of lanes	Saturation Flow rate (pcpchg)	Critical hourly volume (vph)
1	Northbound & Southbound left	10	1	1450	137
2	Northbound & southbound through	25	2	3420	476
3	Eastbound & Westbound left	12	1	1520	173
4	Eastbound & Westbound through	33	3	5145	735

According to Webster's theory, the optimal cycle length (seconds) is
- A. 75
- B. 68
- C. 57
- D. 51

408

A circular horizontal curve has radius, R = 1100 ft and deflection angle between tangents I = 34°45' Rt. If the PI is located at 23 + 12.52, the PT is located at most nearly:
- A. 22 + 16.41
- B. 26 + 35.32
- C. 26 + 56.71
- D. 30 + 75.62

409

A two-lane roadway has a horizontal circular curve (centerline radius 900 ft) as part of its alignment. At the location shown, a tree exists adjacent to the inside lane as shown. Lane width is 12 ft.

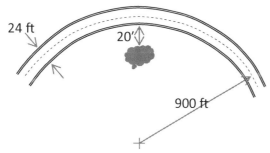

The maximum safe speed (mph) based on adequate stopping sight distance is most nearly:
- A. 45
- B. 50
- C. 55
- D. 60

410

A roadway alignment consists of a circular curve (radius 1780 ft). The maximum permitted superelevation is 8%. If the design speed on the highway is 55 mph, the recommended superelevation (percent) should be most nearly:

    A. 7.8
    B. 6.2
    C. 5.1
    D. 3.5

A Policy on the Geometric Design of Highway and Streets, 6th edition, AASHTO.

411

A parabolic vertical curve is to connect a tangent of +5% to a gradient of –4%. If the PVI is a station 123+32.50 and the tangent offset at the PVT is 17.65 ft, the station of the PVC is most nearly:

    A. 120 + 32.98
    B. 120 + 81.32
    C. 121 + 07.74
    D. 121 + 36.39

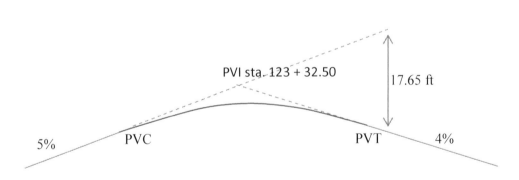

412

A parabolic crest vertical curve connects a grade of +6% to a grade of –4% with the tangents intersecting at station 20 + 23.40 (elevation 245.43 ft above sea level). The design speed on the highway is 60 mph. In order to provide minimum required stopping sight distance, the location of the PVC needs to be most nearly:

    A. 5 + 38.90
    B. 7 + 65.65
    C. 12 + 81.25
    D. 15 + 07.90

413

A parabolic vertical curve connects a downgrade of 5% to an upgrade of 3% as shown. The tangents intersect at station 45 + 12.65 and elevation 234.28 ft above sea level. An overhead structure exists at station 40 + 34.23. If the elevation of the low point on the structure is 275.82 ft, the vertical clearance (feet) at the location is most nearly:

   A. 15.7
   B. 15.2
   C. 14.6
   D. 13.9

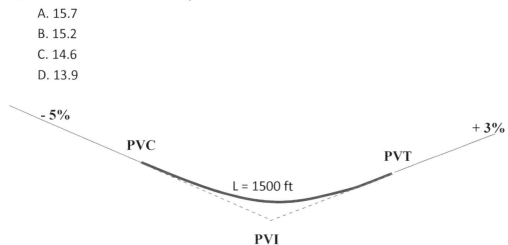

414

The figure shows an unsignalized intersection between Thatcher Lane (2 lanes in each direction) and Fleet Street (one lane in each direction). The design speed on Thatcher Lane is 45 mph. Traffic on Thatcher Lane consists of 6% trucks. Stop signs on both approaches of Fleet Street control access to Thatcher Lane.

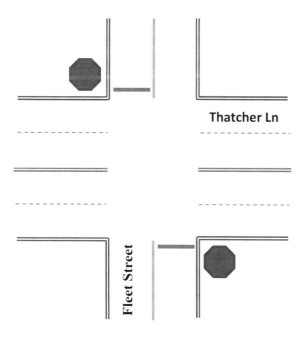

The sight distance (feet) for a stopped passenger car on the northbound approach of Fleet Street to make a left turn onto Thatcher Lane is most nearly:

A. 500
B. 530
C. 565
D. 605

415

The cross-section of a highway (ADT = 5000 vpd) is shown below. The design speed on the highway is 55 mph. At the location of interest, the alignment curves to the left (radius 900 ft). The lateral width X (feet) beyond the outside edge of the 6 ft shoulders is most nearly

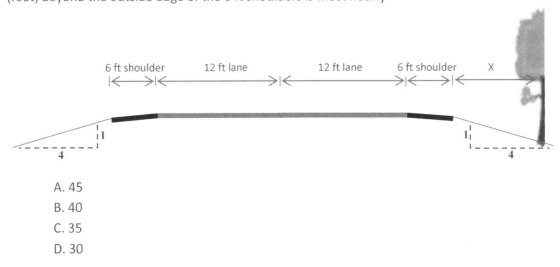

A. 45
B. 40
C. 35
D. 30

416

The minimum width (inches) of the solid white lines marking both edges of pedestrian crosswalks is:

A. 3
B. 4
C. 6
D. 8

Manual of Uniform Traffic Control Devices, 2009

417

A 1-lane major street intersects with a 1-lane minor street at an at-grade intersection. Traffic counts were collected during 12 consecutive hours of a typical day.

	Major Street		Minor Street	
Time	NB	SB	EB	WB
6 AM – 7 AM	410	135	45	55
7 AM – 8 AM	725	320	110	160
8 AM – 9 AM	650	430	100	180
9 AM – 10 AM	570	500	80	100
10 AM – 11 AM	450	530	90	110
11 AM – 12 PM	450	550	80	110
12 PM – 1 PM	550	600	120	110
1 PM – 2 PM	410	620	160	180
2 PM – 3 PM	390	640	130	150
3 PM – 4 PM	350	710	160	190
4 PM – 5 PM	400	750	200	190
5 PM – 6 PM	420	720	170	150

Which of the following statements are true?

    A. Neither warrant 1 nor warrant 2 are met

    B. Both warrants 1 and 2 are met

    C. Warrant 1 is met but warrant 2 is not

    D. Warrant 1 is not met but warrant 2 is met

418

According to the Manual of Uniform Traffic Control Devices, the average walking speed (feet per second) for pedestrians is most nearly:

    A. 3.5

    B. 4.0

    C. 4.5

    D. 5.0

419

A parabolic vertical curve joins a grade of –5% to a grade of +6%. The PVC is at station 53+12.50 and the PVI is at station 60+09.00. Based on rider comfort, what is the maximum recommended speed (mph) on the roadway?

    A. 65

    B. 70

    C. 75

    D. 80

420

A stream of 10 vehicles has their speed measured by recording the time to traverse a 300 ft stretch of a roadway. The speeds (mph) recorded are 45, 54, 47, 62, 48, 44, 50, 45, 44, 60. The average space mean speed (mph) is most nearly:

    A. 51.2
    B. 50.4
    C. 49.9
    D. 49.2

421

Citizens of a township have three choices for commuting to the employment zone defined as 'downtown', which is a distance of 12 miles away. The choices are (1) driving, (2) commuter rail and (3) a suburban bus line. The utility values for the three modes, based on a factored sum of attributes such as time, cost, etc. are -1.23 for driving, -2.11 for commuter rail and -1.89 for the bus line. The total number of commuters originating from the community is 820. The number of driving trips expected is most nearly:

    A. 425
    B. 303
    C. 255
    D. 153

422

Traffic counts were obtained in 15 minute intervals and recorded as shown in the table below.

Time interval	Vehicles
7:00 AM – 7:15 AM	239
7:15 AM – 7:30 AM	312
7:30 AM – 7:45 AM	343
7:45 AM – 8:00 AM	322
8:00 AM – 8:15 AM	307
8:15 AM – 8:30 AM	380
8:30 AM – 8:45 AM	372
8:45 AM – 9:00 AM	345

The peak hour factor is most nearly:

    A. 0.92
    B. 0.86
    C. 1.16
    D. 1.08

423

A two-lane highway has 12 ft wide lanes and carries significant truck traffic. The design vehicle is the WB-62 truck (interstate semitrailer). At a particular location, the alignment consists of a circular curve of radius 1090 ft. If the design speed is 60 mph, the roadway widening (feet) required at this location is most nearly:

    A. 0.0
    B. 2.0
    C. 3.0
    D. 4.0

424

The 48-hour traffic count for an urban arterial results is a count of 20,890 veh/day. The seasonal factor is 0.88 and the axle correction factor is 0.92. If the K value for similar roadways is 9.5%, the design hourly volume (veh/hr) should be most nearly:

    A. 700
    B. 1,250
    C. 1,400
    D. 1,600

425

In the 2010 edition of the Highway Capacity Manual, the pedestrian link LOS score is based on:

    A. pedestrian volume, link gradient and cross section factor
    B. vehicular volume, vehicle speed and cross section factor
    C. vehicular volume, pedestrian volume and vehicle speed
    D. pedestrian volume, vehicular volume and link gradient

426

The data from a weigh station consists of axle load data from 125 trucks. All axles fell into either single-axle or tandem-axle varieties.

SINGLE AXLES		
Axle Load (kips)	Number	LEF
6	21	0.009
8	56	0.031
12	48	0.176
TANDEM AXLES		
Axle Load (kips)	Number	LEF
14	78	0.024
18	85	0.070
22	90	0.166
26	34	0.342

The truck factor (TF) is most nearly:
- A. 0.11
- B. 0.36
- C. 2.79
- D. 9.20

427

Which of the following statement is/are true for a continuously reinforced concrete pavement?
  I. Increasing pavement thickness typically decreases the density pf punchouts
  II. Punchouts are caused by compressive stresses in the slab due to wheel loads
  III. Pavement durability is affected by the type of base used to support the pavement slab
  IV. Durability of a pavement is directly related to number of load cycles
  V. The MEPDG doe s not allow the designer to design a pavement with a specific upper limit on crack density

- A. All of them
- B. I, III and IV only
- C. I, III, IV and V only
- D. I and IV only

428

The results of a standard Proctor test are summarized in the figure. A sample of soil has volume = 0.3 ft³ and has water content = 15%.

The weight of the soil sample (lb) is most nearly:
   A. 32.1
   B. 30.2
   C. 28.6
   D. 25.9

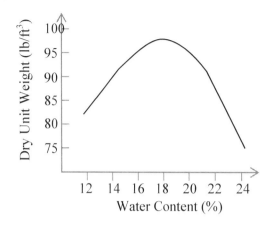

429

The mass diagram shown below represents the earthwork for a highway project. Stations are marked every 100 feet. Which of the following statements is/are true?

   I. Earthwork is balanced between stations 0 + 00 and 12 + 00
   II. Station 5 + 00 represents a transition from a fill to a cut
   III. Station 9 + 00 represents a transition from a fill to a cut

   A. All of them
   B. I and II only
   C. II and III only
   D. I and III only

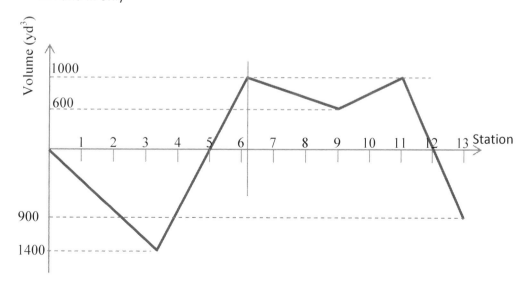

430

A two-lane freeway segment with a single-lane on ramp followed by a single-lane off ramp is shown below. All indicated volumes are hourly volumes. PHF on the freeway is 0.88 and on the ramps is 0.95.

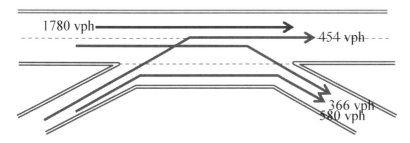

The volume ratio for the weaving segment is most nearly:

A. 0.25
B. 0.35
C. 0.65
D. 0.75

431

A trapezoidal open channel conveys a flow rate = 400,000 gal/min. The Manning's n = 0.015. The bottom width of the channel = 20 ft and longitudinal slope of the channel floor is 0.8%. The side slopes are 2:1

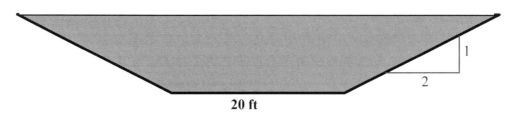

The depth of flow (feet) is most nearly:
A. 5.00
B. 4.25
C. 3.75
D. 2.50

432

A stream collects runoff from a watershed area A = 325 acres. The table below shows discharge (ft³/sec) recorded in the stream during the period t = 0 to t = 5 hours after a rainfall event.

Time (hr)	Flow Rate (ft³/sec)
0	20
1	50
2	70
3	120
4	60
5	20

The average depth of runoff (inches) from the watershed is most nearly:
  A. 0.36
  B. 0.54
  C. 0.67
  D. 1.04

433

Particle size distribution of a soil sample is summarized in the table below. The fines were tested using Atterberg apparatus to obtain the following results:
Liquid limit   45
Plastic limit   21

Sieve size	% finer
1.0 inch (25.4 mm)	100
0.5 inch (12.7 mm)	92
No. 4 (4.75 mm)	75
No. 10 (2.00 mm)	62
No. 40 (0.425 mm)	52
No. 100 (0.15 mm)	45
No. 200 (0.075 mm)	28

The classification of the soil according to the AASHTO Soil Classification system is:
  A. A-1-a
  B. A-4
  C. A-2-5
  D. A-2-7

434

A single-lane roundabout (single entry lane conflicting with a single circulating lane) has a conflicting volume = 230 pc/h. The capacity (pc/h) of the entering lane, adjusted for heavy vehicles is most nearly:
   A. 865
   B. 900
   C. 945
   D. 1010

435

Which of the following is NOT an objective for traffic barriers?
   A. Increasing capacity
   B. Separating opposing flow
   C. Channeling various modes of traffic flow
   D. Work zone safety

436

For a newly constructed curb ramp of length 12 feet, ADA requirements limit the maximum ramp slope (percent) to:
   A. 5.00
   B. 6.67
   C. 8.33
   D. 10.0

437

According to ADA accessibility guidelines, the minimum number of handicapped accessible spaces in a parking lot which accommodates a total of 630 vehicles is most nearly:
   A. 7
   B. 9
   C. 10
   D. 13

438

The figure shows a signalized intersection between Main Street (2 lanes in each direction) and North Avenue (one lane in each direction). The design speed on Main Street is 45 mph and traffic consists of 6% trucks. The signal cycle length for the intersection is 75 seconds. The pedestrian flow rate at the N-S crosswalk is 400 ped/hr and the effective 'walk' time for the minor street through movement is 15 seconds. The minimum green time (seconds) for the north south signal phase based on pedestrian crossing time on the N-S crosswalk is most nearly:

   A. 19
   B. 26
   C. 13
   D. 21

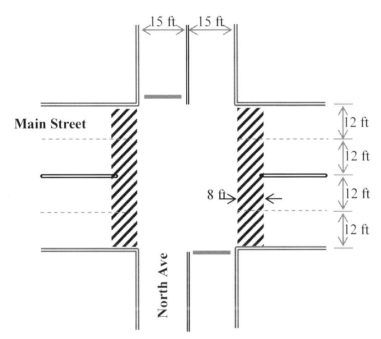

439

According to the Manual of Uniform Traffic Control Devices, 2009, a temporary traffic control designated 'long-term stationary' is one that occupies a fixed work zone for a period exceeding

   A. 3 days
   B. 4 days
   C. 5 days
   D. 7 days

440

A structure has been examined to determine a need for rehabilitation. The related costs are summarized below:

   Current annual costs = $40,000
   Estimated rehabilitation cost = $350,000
   Annual costs projected after rehabilitation = $15,000
   Expected useful life remaining = 20 years
   Projected increase in residual value (at end of useful life) = $200,000

The return on investment (ROI) for performing the rehabilitation is most nearly:
   A. 5%
   B. 6%
   C. 7%
   D. 8%

THIS IS THE END OF THE TRANSPORTATION DEPTH EXAM

SOLUTIONS TO BREADTH EXAM
FOR THE
CIVIL PE EXAM

ANSWER KEY: BREADTH EXAM

001	A
002	D
003	B
004	C
005	C
006	B
007	D
008	A

009	C
010	B
011	C
012	A
013	D
014	B
015	B
016	C

017	D
018	C
019	B
020	A
021	C
022	D
023	B
024	A

025	A
026	D
027	D
028	A
029	B
030	D
031	A
032	B

033	A
034	D
035	A
036	C
037	B
038	C
039	D
040	B

Solution 001

From the table on the right, the total annual count = 225,680, from which we obtain an average monthly ADT = 18,807

Therefore, the monthly expansion factor for April = 18,807÷21,983 = 0.856

From the table on the left, the 7-day count = 123,355, of which 26,485 is weekend traffic. Therefore, the cumulative (Mon-Fri) weekday traffic is 96,870 (5 day average of 19,374.

Therefore, the daily expansion factor for Wednesday = 19,374÷19,882 = 0.974

Therefore the AAWT = 19,545x0.856x0.974 = 16,296

Answer is A

Solution 002

The easiest way to solve this problem is to use the formula for the area of a triangle:
$$A = \frac{1}{2} ab \sin C$$
where C is the angle between two sides a and b
The azimuth angles of the two lines are 65.75 and 102 degrees. Therefore the angle between them is 102 − 65.75 = 36.25
Area: $A = \frac{1}{2} \times 2345 \times 3020 \times \sin 36.25 = 2,093,798\ ft^2 = 48.067\ acres$
Answer is D

Alternatively: Arbitrarily assuming the coordinates of A to be (0, 0), the coordinates of the other two points can be found and then the method of coordinates can be used.

Solution 003

The pipe length between the two manholes has length = 1521.3 − 1062.6 = 458.70 ft.

The point of interest (low ground elevation) is located at a distance = 1305.1 – 1062.6 = 242.50 ft. from the upstream end, and 216.2 ft from the downstream end

The invert elevation at this location can be calculated by averaging the upstream and downstream invert elevations, as follows

$$238.98 \times \frac{216.2}{458.7} + 230.65 \times \frac{242.5}{458.7} = 234.58$$

Pipe wall thickness = (36.3 – 32.2)/2 = 2.05 in = 0.171 ft
Outer diameter = 36.3 in = 3.025
Elevation of the TOP of pipe = 234.58 – 0.171 + 3.025 = 237.43 ft
Soil cover = 241.55 – 237.43 = 4.12 ft

Answer is B

Solution 004

The perimeter of each gutter is 3.162 ft
Total perimeter of gutters = 6.324 ft
Surface area per mile = 6.324x5280 = 33,393.65 ft^2/mile
With a thickness of 3 inches, the volume of concrete = 8,348.4 ft^3/mile = 309.2 yd^3/mile
Cost of concrete material and placement = 232 x 309.2 = $71,734.40

Answer is C

Solution 005

Starting from A: ES_A = 0; EF_A = 0 + 5 = 5. This carries over to the successor C
For activity C: ES_C = 5; EF_C = 5 + 3 = 8
For activity B, there are two predecessors (A and C): ES_B = larger of EF_A and EF_C = 8; EF_B = 8 + 4 = 12
Since D has a single predecessor (B), ES_D = EF_B = 12. Therefore EF_D = 12 + 3 =15
E has two predecessors (B and C). Therefore ES_E = larger of EF_B and EF_C = 12. And EF_E = 12 + 5 = 17
Based on the FF lag between D and F, the EF_F = 18, based on which the ES_F = 16. However, based on EF_E = 17, the ES_F = 17. This controls.

Answer is C

Solution 006

The elevation of the point on the curve is 2.5 ft (30 in) above the crown of the sewer pipe, therefore at elev. 305.15. This point has the following offsets from the PVI:

horizontal offset h = 1230.05 – 1145.20 = +84.85 ft

vertical offset v = 305.15 – 310.56 = - 5.41 ft

The maximum length of curve (L) can be calculated from

$$\frac{L+2h}{L-2h} = \sqrt{\frac{v-G_1h}{v-G_2h}} = \sqrt{\frac{-5.41-0.05\times84.85}{-5.41--0.03\times84.85}} = \sqrt{\frac{-9.6525}{-2.8645}} = 1.836$$

$$\frac{L+169.7}{L-169.7} = 1.836$$

Solving L = 575.7 ft. Since this is the maximum length of curve, look for the next lower value

Answer is B

Solution 007

For ϕ = 32, active earth pressure coefficient: $K_a = \frac{1-\sin\phi}{1+\sin\phi} = 0.307$

At the base of the footing (depth = 18 ft), the effective earth pressure: $K_a \gamma_{sub} H = 0.307 \times (126 - 62.4) \times 18 = 351.5\ psf$
At the base of the footing (depth = 18 ft), the hydrostatic pressure: $\gamma_w H = 62.4 \times 18 = 1123.2\ psf$
Total pressure at bottom of footing = 1474.7 psf

Total active resultant = 0.5x1474.7x18 = 13,272 lb/ft

Answer is D

Solution 008

Initial effective stress at midheight of clay layer (17 ft below surface): $p_1' = 5 \times 115 + 7 \times (125 - 62.4) + 5 \times (125 - 62.4) = 1326.2\ psf$
After lowering water table, effective stress: $p_2' = 12 \times 115 + 5 \times (125 - 62.4) = 1693\ psf$
Ultimate consolidation settlement: $s = \frac{HC_c}{1+e_o}\log_{10}\frac{p_2'}{p_1'} = \frac{120\times0.46}{1+0.45}\log_{10}\frac{1693}{1326.2} = 4.04\ in$

Time t = 3 months = 90 days; Drainage path H_d = 5 ft (double drainage)
From the settlement time relationship, the time factor is: $t = \frac{T_v H_d^2}{c_v} \Rightarrow T_v = \frac{c_v t}{H_d^2} = \frac{0.01\times90}{5^2} = 0.036$

Corresponding degree of consolidation = 22%

Therefore, after 3 months, settlement = 0.22X4.04 = 0.89 in

Answer is A

Solution 009

There are 4 loading zones on the beam – the load function on them, left to right, are w = 0, w = constant, w = 0 and w = 0 respectively. As a result, the bending moment function is M = linear, quadratic, linear and linear respectively. This eliminates choices A, B and D.

Answer is C

Solution 010

The maximum compressive stress will occur at the upper left corner of the cross section, where the uniform compression P/A will combine with the bending stress components produced by the moment about either axis (Mc/I)

$$\sigma = \frac{20 \times 10^3}{0.2 \times 0.25} + \frac{20 \times 10^3 \times 0.07 \times 0.125}{\frac{1}{12} \times 0.2 \times 0.25^3} + \frac{20 \times 10^3 \times 0.04 \times 0.1}{\frac{1}{12} \times 0.25 \times 0.2^3} = 1.552 \times 10^6 \, Pa$$

Answer is B

Solution 011

With 2:1 side slopes and a depth of 5 ft, the width at the top surface = 20 + 2x10 = 40 ft
Area of flow, A = 150 ft²
Wetted perimeter, P = 20 + 2x5x√5 = 42.36 ft
Hydraulic radius, R_h = 150/42.36 = 3.54 ft

Velocity: $V = \frac{1.486}{0.015} \times 3.54^{2/3} \times \sqrt{0.008} = 20.58 \, fps$

Flow rate: Q = VA = 20.58X150 = 3087.2 cfs = 1995.2 MGD

Answer is C

Solution 012

Ultimate bearing capacity of a square footing, according to Terzaghi's theory, is given by
$$q_{ult} = 1.3cN_c + \gamma D N_q + 0.4\gamma B N_\gamma$$
From the supplied figure, for ϕ = 30°, N_c = 30, N_q = 18.5, N_γ = 22.5
$$q_{ult} = 1.3 \times 200 \times 30 + 120 \times 3 \times 18.5 + 0.4 \times 120 \times 5 \times 22.5 = 19,860 \, psf$$

Soil pressure at base of footing = column load + soil overburden = 140,000/25 + 120X3 = 5960 psf

FS = 19,860/5,960 = 3.33

Answer is A

Solution 013

Using a tributary width of 28 inches (2.33 ft) for each stud, the uniform load acting on each stud is 30x2.33 = 70 lb/ft

Maximum bending moment in stud: $M = \frac{wL^2}{8} = \frac{70 \times 10^2}{8} = 875 \, lb \cdot ft = 10{,}500 \, lb \cdot in$

Section modulus of stud (about major axis): $S = \frac{bh^2}{6} = \frac{1.5 \times 5.5^2}{6} = 7.56 \, in^3$

Maximum bending stress: $\sigma = \frac{M}{S} = \frac{10500}{7.56} = 1389 \, psi$

Answer is D

Solution 014

In the table below, cut and fill volumes between stations are calculated using the average end area method. The net volume is calculated in the last column (positive for cut and negative for fill). The cumulative earthwork volume is the sum of the numbers in the last column.

Station	Volume (yd³) CUT	Volume (yd³) FILL	Net Volume (yd³)
0 + 0.00			
	1032.4	370.0	+ 662.4
1 + 0.00			
	1340.7	141.3	+ 1199.4
2 + 0.00			
	1196.3	163.7	+ 1032.6
3 + 0.00			
	1445.7	560.9	+ 884.8
4 + 0.00			
			+ 3779.2

Answer is B

Solution 015

Maximum dry unit weight (Proctor) = 109 pcf
Therefore, required dry unit weight = 0.9x109 = 98.1 pcf
Weight of soil solids in embankment = 98.1x50,000 = 4.905x10⁶ lb
Dry unit weight of borrow soil = 120 ÷ 1.14 = 105.3 pcf
Therefore, volume of borrow soil needed = 4.905x10⁶ ÷ 105.3 = 46,598 ft³ = 1,725.8 yd³

Answer is B

Solution 016

The table below shows the precipitation in each of the three regions and the calculation of the weighted average as the average precipitation over the entire (268 acre) area

Region between	Area enclosed (acres)	Average Precipitation (inches)
1 & 2	147	0.4
2 & 3	76	0.75
3	45	0.9

The weighted average is the calculated as:

$$\bar{P} = \frac{\sum P_i A_i}{\sum A_i} = \frac{0.4 \times 147 + 0.75 \times 76 + 0.9 \times 45}{147 + 76 + 45} = 0.58 \; in$$

Answer is C

Solution 017

Using a Hazen Williams roughness coefficient C = 100 and a length L= 1 mile =5280 ft

$$h_f = \frac{4.725 Q_{cfs}^{1.85} L_{ft}}{C^{1.85} D_{ft}^{4.865}} = \frac{4.725 \times 12.5^{1.85} \times 5280}{100^{1.85} \times 2^{4.865}} = 18.3 \; ft$$

Answer is D

Solution 018

Volume of pond to be filled before it starts emptying = 0.78x760,000 = 592,800 gal = 79,251 ft³
At the rate of 2 cfs, this will require 39,626 seconds = 11 hours

Answer is C

Solution 019

Taking moments about support A,

$$\sum M_A = 20 \times 10 + 10 \times 14 + 40 \times 15 + 60 \times 30 + 40 \times 45 - 60 E_y = 0 \Rightarrow E_y = 75.67$$

Making a section through CD, DG and FG and then taking moments about G (see free body diagram below)

$$\sum M_G = 40 \times 15 - 75.67 \times 30 - \frac{4}{15.52} F_{CD} \times 15 - \frac{15}{15.52} F_{CD} \times 10 = 0 \Rightarrow F_{CD} = -123.43$$

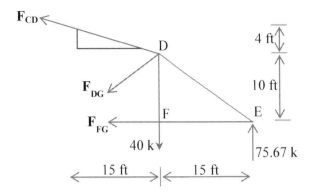

Answer is B

Solution 020

Factor of safety is given by:
$$FS = \frac{c}{\gamma H \cos^2 \beta \tan \beta} + \frac{\tan \phi}{\tan \beta}$$

For the riprap layer:
$$FS = 0 + \frac{\tan 36}{\tan 30} = 1.26$$

For the soil:
$$FS = \frac{600}{125 \times 20 \times \cos^2 30 \tan 30} + \frac{\tan 28}{\tan 30} = 0.55 + 0.92 = 1.47$$

The governing FS is 1.26

Answer is A

Solution 021

The fines fraction F_{200} = 28%. Therefore (since F_{200} < 50) the soil is predominantly coarse grained. First letter is S or G. Of the coarse fraction (72%), less than half (25%) is coarser than a no. 4 sieve. Therefore, the first letter is S. This eliminates A and D.

Since F_{200} > 12%, second letter of the classification is determined entirely by plasticity characteristics.
PI = 45 − 21 = 24. LL = 45 and PI = 24.
This plots above the A-line. So, second letter is C.

Answer is C

Solution 022

Mass of water = 1685 − 1418 = 267 g
Volume of soil solids, V_s = 1418 ÷ 2.65 = 535 cc
Therefore, volume of voids, V_v = 855 − 535 = 320 cc
Void ratio e = V_v/V_s = 320/535 = 0.6
Answer is D

Solution 023

Buckling about the weak axis is prevented because of the bracing provided by the nails.
The Euler buckling load (about the strong axis) is given by:
$$P_e = \frac{\pi^2 EI}{L^2} = \frac{\pi^2 \times 1.5 \times 10^6 \times \frac{1}{12} \times 1.5 \times 5.5^3}{(12 \times 12)^2} = 14848 \; lb = 14.85 \; kips$$
Answer is B

Solution 024

Using the reservoir surface as point 1 and open end of pipe as point 2, both of these points are at atmospheric pressure $p_1 = p_2 = p_{atm}$. Also, since the reservoir is 'large', by the continuity principle, $V_1 \approx 0$

Head loss (friction) in the pipe, using the Darcy-Weisbach equation, is
$$h_f = f \frac{L}{D} \frac{V^2}{2g} = 0.024 \times \frac{500}{2} \times \frac{V^2}{2 \times 32.2} = 0.0932 V^2$$
Writing Bernoulli's equation between points 1 and 2
$$\frac{p_{atm}}{\gamma} + 324.5 + 0 - 0.0932 V^2 = \frac{p_{atm}}{\gamma} + 295.8 + \frac{V^2}{2 \times 32.2} \Rightarrow 0.109 V^2 = 28.7 \Rightarrow V = 16.2 \; fps$$
Flow rate: Q = VA = 50.98 cfs

Answer is A

Solution 025

Time of overland flow = $t_s + t_d$ = 5 + 13 = 18 minutes
Channel travel time = 1200 ÷ 5 = 240 seconds = 4 minutes
Time of concentration for point B = 18 + 4 = 22 min
For duration = 22 min and return period = 20 years, intensity = 1.5 in/hr

Answer is A

Solution 026

The tangent length (calculated from PI, PC pair – can also be calculated from PI-PT pair)

$$T = \sqrt{(1250-100)^2 + (320-817)^2} = 1252.8$$

The azimuth angle of the back tangent (calculated as inverse tan of departure divided by latitude)

$$Az_{BT} = \tan^{-1}\left(\frac{817-320}{100-1250}\right) = 156.63$$

The azimuth angle of the forward tangent (calculated as inverse tan of departure divided by latitude)

$$Az_{FT} = \tan^{-1}\left(\frac{900-817}{1350-100}\right) = 3.80$$

Therefore, the deflection angle I = 3.80 – 156.63 = – 152.83 (negative sign means deflecting left)

Since T = R tan (I/2), solving for R = 302.724 ft

Degree of curve D = 5729.578/R = 18.93 degrees

Answer is D

Solution 027

Height of warehouse = 26 ft
Perimeter = 2x(40+110) = 300 ft
Surface area of 4 walls = 300x26 = 7800 sq. ft.
Inclined length of roof = 2√(8^2 + 20^2) = 43.08 ft
Surface area of roof = 110x43.08 = 4739 sq. ft.
Area of 4 openings to be subtracted = 4x8x13 = 416 sq. ft.
Total area of sheathing = 7800 + 4739 – 416 = 12,123 sq. ft.

Answer is D

Solution 028

Collapsing the 3D structures into a 2D one, and representing the *total* pad reaction on the left (2 pads) as R_L and the *total* pad reaction on the right (2 pads) as R_R, taking moments about R_L, we get:

$$9000 \times 14 + 800 \times 37 - R_R \times 36 + 8000 \times 56 = 0 \Rightarrow R_R = 16,767\ lb$$

With the load on the right, the maximum compression reaction on the ground will occur on the pads on the right.
Reaction on each pad = 8,384 lb

Soil pressure under the pads on the right (area = 7.07 ft^2) = 1,186 lb/ft^2

Answer is A

Solution 029

For concrete beams, the bending moment due to loading determines the extent of cracking experienced by the beam. The effective moment of inertia, used for calculating deflections according to the elastic theory, is between I_{cr}, the cracked moment of inertia and I_g, the gross moment of inertia. Only statement IV is correct.

Answer is B.

Solution 030

By using the 0.2% offset method (drawing a line parallel to the initial tangent through the strain offset = 0.2% = 2000 µε), the yield stress is 38 ksi

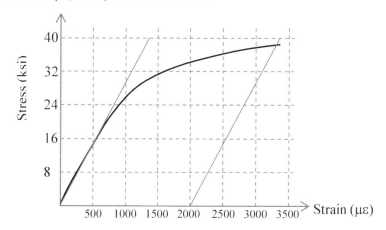

Answer is D

Solution 031

The thickness of sand that is affected by the pressure from the mat foundation is 15 – 3 = 12 ft = 180 in
Assuming that the mat exerts a uniform pressure of 600 psf on the entire layer,
Vertical strain = Vertical stress/E = $600/1 \times 10^6$ = 6×10^{-4}
Vertical displacement = $180 \times 6 \times 10^{-4}$ = 0.108 in

Answer is A

Solution 032

II is incorrect. LS – EF – D = LS – (ES + D) – D = LS – ES – 2D = TF – 2D, which can equal TF only if duration D = 0 (which is a trivial solution)
Answer is B

Solution 033

In the long term, clay soils go through consolidation due to the expulsion of pore water. Results obtained from the CD (consolidated drained) triaxial test are good predictors of long term conditions. Therefore, statement I is correct.
The CD test involved allowing pore water to slowly drain from the soil and is therefore slower than the UU test. Statement II is correct and statement III is false. During the CD test, pore pressures are not allowed to build up. Statement IV is incorrect.
Answer is A

Solution 034

Material delivery = 540 yd^3/hr (loose soil), which is equivalent to 540x0.83 = 448.2 yd^3/hr compacted
Roller covers ground at 3 mph x 8 ft = 126720 ft^2/hr. 0.5 ft thick layer gets compacted in 4 passes. Therefore each pass compact the equivalent of 0.125 ft, which means it compacts 15,840 ft^3 (587 yd^3) of soil per pass. This is ideal capacity. Working 50 minutes per hour, roller compacts 50/60x587 = 489 yd^3/hr. Therefore, only 1 roller is needed to handle the delivery of the material.

Answer is D

Solution 035
Midspan deflection of a simply supported beam with point load is given by:
$$\Delta_{max} = \frac{PL^3}{48EI} = \frac{20 \times (24 \times 12)^3}{48 \times 29000 \times 1490} = 0.23\ in$$
Answer is A

Solution 036

The total vertical force (weight of concrete 3 & 4) + weight of soil (1) = 20,600 lb/ft
Horizontal friction force that can be mobilized under the footing = 0.6x20,600 = 12,360 lb/ft
$$FS = \frac{12,360}{5,400} = 2.29$$
Answer is C

Solution 037

The total volume of runoff that collects in the detention pond = 120 acre x 1 inch = 120 ac-in = 4.356x10^5 ft^3 = 12,333 m^3 = 1.2x10^7 L
Total mass of sediment = 6.167x10^7 g = 6.167x10^4 kg
Bulk specific gravity of the sediment = 80/62.4 = 1.282
Density of sediment = 1282 kg/m^3
Volume occupied by sediment = 6.167x10^4 kg ÷ 1282 kg/m^3 = 48.1 m^3 = 1699 ft^3
Depth occupied by sediment = 1699 ÷ (2x43560) = 0.0195 ft = 0.234 inch
Answer is B

Solution 038

Breaking strain calculated as: Elongation at break ÷ Sample initial length
The values calculated for samples 1-5 are: 0.0728, 0.0703, 0.0719, 0.0716 and 0.0709
The average breaking strain = 0.0715 = 71,500 µε

Answer is C

Solution 039

Depth ratio d/D = 30/48 = 0.625. For this depth ratio, for constant n, $V/V_f = 1.0857$

Longitudinal slope: $S = \frac{365.82-354.28}{1200} = 0.0096$

Velocity for pipe flowing full: $V_f = \frac{0.590}{n}D^{2/3}S^{1/2} = \frac{0.590}{0.014} \times 4^{2/3} \times 0.0096^{1/2} = 10.4 \; cfs$

Actual velocity in pipe (when d = 30 in) is V = 11.3 fps

Answer is D

Solution 040

Underpinning (I) is commonly used to support existing structures to counter the possible loss of bearing support from adjacent excavation. Slurry walls (III) can be used to 'isolate' sensitive structures from construction activities.

Answer is B

SOLUTIONS TO TRANSPORTATION EXAM
FOR THE
CIVIL PE EXAM

ANSWER KEY: TRANSPORTATION DEPTH EXAM

401	D
402	C
403	D
404	A
405	C
406	D
407	C
408	B

409	B
410	B
411	D
412	C
413	A
414	C
415	B
416	C

417	B
418	A
419	C
420	D
421	A
422	A
423	C
424	D

425	B
426	B
427	B
428	A
429	D
430	A
431	D
432	C

433	D
434	B
435	A
436	C
437	D
438	A
439	A
440	B

Solution 401

Assuming level terrain (mild slope, short length), $E_T = 1.5$

Heavy vehicle factor: $f_{HV} = \frac{1}{1+P_T(E_T-1)} = \frac{1}{1+0.08\times(1.5-1)} = 0.962$

Flow rate: $v_p = \frac{V}{PHF f_{HV} f_p N} = \frac{4700}{0.92\times 0.962\times 1.0\times 3} = 1770\ pcphpl$

For 11 ft. wide lanes, $f_{LW} = 1.9$ mph

For 6 ft shoulder, $f_{LC} = 0.0$

Each full cloverleaf has 4 ramps (2 on and 2 off) in each direction

Total ramp density, TRD = 4x4/6 = 2.67 ramps per mile

Free flow speed: $FFS = 75.4 - f_{LW} - f_{LC} - 3.22 TRD^{0.84} = 75.4 - 1.9 - 0 - 3.22\times 2.67^{0.84} = 66.2\ mph$

Using FFS = 65 mph and v_p = 1770, LOS = D

Answer is D

Solution 402

ADT in 2010 = 15,600x1.03^2 = 16,550

Crashes expected (without crash modifications) in 2010 = 12x1.03^2 = 12.73

Overall crash reduction factor: $CR = CR_1 + (1-CR_1)CR_2 + (1-CR_1)(1-CR_2)CR_3 = 0.1 + 0.9\times 0.25 + 0.9\times 0.75\times 0.2 = 0.46$

Number of crashes prevented = 12x0.46x16550/15600 = 5.86

Number of crashes expected in 2010 = 12.73 − 5.86 = 6.87

Answer is C

Solution 403

Total number of observations = 88

85% of observations = 74.8
Speed of 45 mph has an associated cumulative frequency = 74, and
Speed of 50 mph has an associated cumulative frequency = 85

Interpolating, we have 85th percentile speed (corresponding to 74.8 observations) = 45.4 mph

Answer is D

Solution 404

The accident rate for a highway segment is calculated as accidents per hundred million vehicle miles:

$$Rate = \frac{N \times 10^8}{365 \times ADT \times N_{yrs} \times L_{mi}}$$

The values, for segments 1-4, are 114, 240, 101 and 142 respectively. Thus, segments 2 and 4 are the two worst performing.

Answer is A

Solution 405

For acceleration phase, time = 70÷8 = 8.75 sec
Acceleration rate = 8 mph/sec = 11.76 ft/s^2
Distance = ½ at^2 = 0.5x11.76x8.75^2 = 450.2 ft
For deceleration phase, time = 70÷10 = 7.0 sec
Deceleration rate = 10 mph/sec = 14.7 ft/s^2
Distance = ½ at^2 = 0.5x14.7x7^2 = 360.2 ft

Since total distance traveled = 0.5 mile = 2640 ft, this leaves distance for constant velocity phase = 2640 − 450.2 − 360.2 = 1829.6 ft

Time for constant velocity phase = 1829.6÷(1.47x70) = 17.8 sec

Total travel time = 8.75 + 17.8 + 7.0 = 33.53 sec

Average running speed = 2640 ÷ 33.53 = 78.73 fps = 53.6 mph

Answer is C

Solution 406

X = 1/1.7 = 0.588; Y = 3.7; Z = 134
Daily trips per household: $T = 2.3 \times 0.588 + 0.6 \times 3.7 + 0.02 \times 134 = 6.25$

Total daily trips for the community = 6.25x250 = 1563

Answer is D

Solution 407

The v/s ratio for phases 1-4 can be calculated as:
Phase 1: v/s = 137 ÷ 1450 = 0.0945
Phase 2: v/s = 476 ÷ 3420 = 0.1392
Phase 3: v/s = 173 ÷ 1520 = 0.1138
Phase 4: v/s = 735 ÷ 5145 = 0.1429

Total lost time = 4x4 = 16 seconds

Webster's optimal cycle length: $C = \frac{1.5L+5}{1-\Sigma\left(\frac{v}{s}\right)_i} = \frac{1.5 \times 16+5}{1-0.4904} = 57\ sec$

Answer is C

Solution 408

Degree of curve: $D = \frac{5729.578}{R} = 5.21°$

Length of curve: $L = 100\frac{I}{D} = 100 \times \frac{34.75}{5.21} = 666.99\ ft$

Tangent length: $T = R \tan\left(\frac{I}{2}\right) = 1100 \times \tan\left(\frac{34.75}{2}\right) = 344.19\ ft$

sta. PC (feet) = sta. PI (feet) – T = 2312.52 – 344.19 = 1968.33
sta. PT (feet) = sta. PC (feet) + L = 1968.33 + 666.99 = 2635.32 (station 26 + 35.32)

Answer is B

Solution 409

The driver on the inside lane has the greatest obstruction to the line of sight. Assuming that the driver is located on the centerline of the inside lane, this is a circular path with radius = 900 – 6 = 894 ft
The lateral offset between this path and the tree is 20 + 6 = 26 ft.

Using R = 894 ft and M = 26 ft, the stopping sight distance S is given by:

$$S = \frac{R}{28.65}\cos^{-1}\left(1-\frac{M}{R}\right) = \frac{894}{28.65}\cos^{-1}\left(1-\frac{26}{894}\right) = 432\ ft$$

On level ground, a stopping sight distance S = 432 ft corresponds to a speed V = 50.6 ft

Answer is B

Solution 410

According to Figure 3-11 of the AASHTO GDHS 2011, for e_{max} = 8%, R = 1780 ft and V = 55 mph, the recommended superelevation rate is 6.1%

Answer is B

Solution 411

The tangent offset at any location on a vertical curve = ½ Rx²

At the end of the curve (i.e. at the PVT) x = L, therefore tangent offset = $\frac{1}{2}\frac{G_2-G_1}{L}L^2 = \frac{(G_2-G_1)L}{2}$

For a crest curve, the vertical offset is negative, therefore: $\frac{(-4-5)L}{2} = -17.65 \Rightarrow L = 3.9222\ sta.$

Therefore, since the PVC is half the curve length upstream of the PVI:

Sta. PVC = sta. PVI − 1.9611 = 123.325 − 1.9611 = 121.3639   **(121 + 36.39)**

Answer is D

Solution 412

Algebraic difference in grades, A = 10%

For design speed = 60 mph on level grade (using AASHTO default values for reaction time and deceleration rate), stopping sight distance S = 566 ft

For the crest curve:

For S ≤ L, $L = \frac{AS^2}{2158} = \frac{10 \times 566^2}{2158} = 1484.5$

This solution satisfies the condition S < L. Therefore, it is OK

PVI is located at 20 + 23.40. Therefore PVC is located half the curve length behind the PVI

PVC location at 2023.40 − 742.25 = 1281.25 (station 12 + 81.25)

Answer is C

Solution 413

Rate of change of gradient: $R = \frac{G_2-G_1}{L} = \frac{3-(-5)}{15} = +0{,}533\ \%/sta$

PVC is located L/2 = 750 ft behind the PVI = 4512.65 − 750 = 3762.65 ft
Distance from PVC to location of structure = 4034.23 − 3762.65 = 271.58 ft (2.7158 stations)

Elevation of PVC = elevation of PVI − G_1 (L/2) = 234.28 − (−5x7.5) = 271.78

Elevation at station 40 + 34.23 is:

$$y = y_{PVC} + G_1 x + \frac{1}{2} R x^2 = 271.78 + (-5)(2.7158) + \frac{1}{2}(0.533)(2.7158)^2 = 260.17 \; ft$$

Vertical clearance = 275.82 − 260.17 = 15.65 ft

Answer is A

Solution 414

The maneuver described (left turn from stop on minor) is Case B1 (Intersection Sight Distance in AASHTO Green Book). For a passenger car to make this maneuver, the minimum time gap t_g = 8.0 sec. For the additional lane to cross, add 0.5 sec to t_g. Therefore, the time gap is 8.5 sec

$$ISD = 1.47 V t_g = 1.47 \times 45 \times 8.5 = 562 \; ft$$

Answer is C

Solution 415

From the AASHTO Roadside Design Guide, for 4:1 fill slope and ADT = 5000 and a design speed of 55 mph, the clear zone distance L_c = 30 ft

For a curve radius of 900 ft and design speed = 55 mph, the curve adjustment factor K_{cz} = 1.5

Therefore, required clear zone width = 30x1.5 = 45 ft. Since the shoulder is part of the recovery area, the width X = 45 – 6 = 39 ft

Answer is B

Solution 416

Although the MUTCD does permit some variations for additional visibility, the basic specifications call for solid white lines not less than 6 inch wide marking both edges of the crosswalk and spaced at least 72 inches apart (MUTCD 2009 Section 3B.18).

Answer is C

Solution 417

WARRANT 1: EIGHT-HOUR VEHICULAR VOLUME

Warrant 1: Condition A thresholds (100% level) for 1 lane major – 1 lane minor are 500 vph total on both approaches major AND 150 vph on higher volume minor. These thresholds are exceeded during only 7 hours. So condition A is NOT MET at 100% level

Warrant 1: Condition B thresholds (100% level) for 1 lane major – 1 lane minor are 750 vph total on both approaches major AND 75 vph on higher volume minor. The two criteria for Condition B of Warrant 1 are met during 11 hours.

Warrant 1 requires the thresholds met during any 8 hours. So, this condition of warrant 1 is met.

Overall, Warrant 1 is met if EITHER condition A or condition B is met at the 100% level

WARRANT 2: FOUR-HOUR VEHICULAR VOLUME

The data in the table is plotted on Figure 4C-1. All but 1 data point plots above the line for 1 lane major &1 lane minor. The warrant requires that this threshold be exceeded for at least 4 hours on a single day. Therefore, this warrant is met.

Answer is B

Solution 418

According to section 4E.06 of the MUTCD, walking speed of pedestrians is to be taken as 3.5 fps

Answer is A

Solution 419

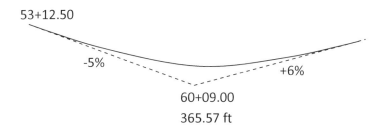

Distance from PVC to PVI = 6009.0 – 5312.50 = 696.50 ft
Length of curve = 2 x 696.50 = 1393 ft

Algebraic difference in grades: A = 6 + 5 = 11%

Minimum length of curve (based on rider comfort): $L_{min} = \frac{AV^2}{46.5}$

Therefore, max speed: $V = \sqrt{\frac{46.5L}{A}} = \sqrt{\frac{46.5 \times 1393}{11}} = 76.7 \; mph$

Answer is C

Solution 420

The space mean speed is calculated using the harmonic mean

$$\bar{V} = \frac{N}{\sum_1^N \frac{1}{V_i}} = \frac{10}{1/45 + 1/54 + \cdots + 1/60} = 49.2$$

Answer is D

Solution 421

The probability of selecting mode A (driving) can be written:

$$P(A) = \frac{e^{U_A}}{e^{U_A} + e^{U_B} + e^{U_C}} = \frac{e^{-1.23}}{e^{-1.23} + e^{-2.11} + e^{-1.89}} = \frac{0.292}{0.292 + 0.121 + 0.151} = 0.518$$

Number of commuters expected to drive = 0.518×820 = 424.5

Answer is A

Solution 422

Counting the five contiguous hours that exist within the 2 hour data, we have the following table:

Time interval	Vehicles
7:00 AM – 8:00 AM	1216
7:15 AM – 8:15 AM	1284
7:30 AM – 8:30 AM	1352
7:45 AM – 8:45 AM	1381
8:00 AM – 9:00 AM	1404

Peak hourly volume V = 1404 during the hour 8:00 - 9:00. Within that peak hour, maximum 15 minute count, v_{15} = 380

Peak hour factor: $PHF = \frac{V}{4v_{15}} = \frac{1404}{4 \times 380} = 0.924$

Answer is A

Solution 423

According to the AASHTO Green Book, for a roadway width = 24 ft, design speed = 60 mph and curve radius = 1090 ft (interpolating between 1000 and 1200 ft), the recommended traveled way widening for a WB-62 vehicle is 2.78 ft

Answer is C

Solution 424

ADT = 20,890 vpd

AADT = ADT x seasonal factor x axle correction factor = 20,890 x 0.88 x 0.92 = 16,913

DHV = K x AADT = 0.095 x 16,913 = 1,607 veh/hr

Answer is D

Solution 425

Answer is B

Solution 426

For each axle category, calculate the product NxLEF, as shown in the table below

SINGLE AXLES			
Axle Load (kips)	Number	LEF	N x LEF
6	21	0.009	0.189
8	56	0.031	1.736
12	48	0.176	8.448
		TOTAL	10.373
TANDEM AXLES			
Axle Load (kips)	Number	LEF	N x LEF
14	78	0.024	1.872
18	85	0.070	5.950
22	90	0.166	14.940
26	34	0.342	11.628
		TOTAL	34.390

Therefore ESAL = 10.372 + 34.390 = 44.763

Truck factor: $TF = \frac{ESAL}{N_{trucks}} = \frac{44.763}{125} = 0.358$

Answer is B

Solution 427

I is a true statement. The number of punchouts per mile, a predictor of pavement performance is very sensitive to pavement thickness. In some cases, a reduction of only ¼ inch in PCC thickness results in a near-doubling of the CRCP punchouts.

II is not true. Punchouts are caused by cyclic TENSILE stresses between the wheels in the upper layers of the pavement.

III is a true statement. The base type selected for support in a CRCP is a critical factor impacting projected performance both in the development of cracks and tight crack widths as well as in resisting foundation layer erosion from repeated loading

IV is a true statement. The MEPDG uses a fatigue model to model the effect of repeated load cycles on the development of cracks

V is not true. The iterative nature of the MEPDG procedure allows the design engineer limits the allowable number of punchouts at the end of the design life to an acceptable level (typically between 10 and 20 per mile) at a given level of reliability.

Answer is B

Solution 428

At w = 0.15, the dry unit weight, γ_d = 93 lb/ft^3
Total unit weight $\gamma = \gamma_d(1+w)$ = 106.95 lb/ft^3
Soil sample weight = 0.3x106.95 = 32.09 lb

Answer is A

Solution 429

Statement I is true. Stations 6+20 and 11+00 have the same ordinate (1000 yd^3), implying that the net earthwork between these stations is zero

Statement II is not true. Station 5+00 has a zero ordinate, implying that the net earthwork between station 0+00 and 5+00 is zero

Statement III is true. Between stations 6+20 and 9+00, the mass diagram goes down, implying a fill, while between stations 9+00 and 11+00, the mass diagram goes up, implying a cut

Answer is D

Solution 430

Converting all hourly volumes to flow rates, we have:

$$v_{FF} = \frac{V_{FF}}{PHF} = \frac{1780}{0.88} = 2023$$

$$v_{FR} = \frac{V_{FR}}{PHF} = \frac{366}{0.88} = 416$$

$$v_{RF} = \frac{V_{RF}}{PHF} = \frac{454}{0.95} = 478$$

$$v_{RR} = \frac{V_{RR}}{PHF} = \frac{580}{0.95} = 611$$

Weaving demand flow rate: $v_W = v_{FR} + v_{RF} = 894$

Non-weaving demand flow rate: $v_{NW} = v_{FF} + v_{RR} = 2634$

Volume ratio: $VR = \dfrac{v_W}{v_W + v_{NW}} = \dfrac{894}{894 + 2634} = 0.253$

Answer is A

Solution 431

Flow rate Q = 400,000 gpm = 200,000 ÷ 448.8 = 891.2 cfs

Instead of using a trial and error solution, we can use a flow parameter K, which is defined as

$$K = \dfrac{Qn}{kb^{8/3}S^{1/2}} = \dfrac{891.2 \times 0.015}{1.486 \times 20^{8/3} \times 0.008^{1/2}} = 0.0341$$

From tables for K (All-In-One Chapter 303 or CERM Chapter 19), for side slope parameter m = 2 and K = 0.0341, the depth ratio d/b = 0.1255

Therefore, the depth of flow, d = 0.1255×20 = 2.51 ft

Answer is D

Solution 432

The first step is to separate the base flow in the stream hydrograph.

Time (hr)	Flow Rate (ft³/sec)	Net flow rate (ft³/sec)
0	20	0
1	50	30
2	70	50
3	120	100
4	60	40
5	20	0
	TOTAL	220

The area under this (Q vs t) curve represents the total volume of excess precipitation (runoff) and is calculated as:

$V = 3600 \times 220 = 792,000 \: ft^3$

Average depth of runoff: $d = \dfrac{V}{A} = \dfrac{792,000}{325 \times 43560} = 0.056 \: ft = 0.67 \: in$

Answer is C

Solution 433

The fines fraction F$_{200}$ = 28%. Therefore (since F$_{200}$ < 35), groups A-4 through A-7 are eliminated.
In order to meet criteria for A-1-a, F$_{200}$ must not exceed 15.
In order to meet criteria for A-1-b, F$_{200}$ must not exceed 25.
In order to meet criteria for A-3, F$_{200}$ must not exceed 10.

None of these criteria are met. Therefore, the soil must be in one of the transition groups.

Since LL =45 > 40 and PI = 24 > 10, the soil is classified as A-2-7

Answer is D

Solution 434

According to the HC 2010, the lane capacity (pc/h) is given by:

$$c_{c,pce} = 1130e^{-1.0\times10^{-3}v_{c,pce}} = 1130 \times e^{-0.001\times230} = 898$$

Answer is B

Solution 435

Traffic barriers are sometimes used to separate bicycles and other slower moving modes from vehicular traffic streams. They are also used to separate opposing directions of vehicular traffic and to provide a safe haven to workers and equipment in a work zone adjacent to traffic lanes.

Even though the use of barriers may result in more streamlined traffic movement and therefore, capacity increases, their design is NOT ostensibly based on that objective (choice A).

Answer is A

Solution 436

For new construction, curb ramp slope should not exceed 8.33%. If this is not feasible, a 12.5% slope is permitted for a 2 foot long ramp, or a 10% slope is permitted for a 5 foot long ramp.

Answer is C

Solution 437

According to guidelines in the Americans with Disabilities Act Accessibility Guidelines, when the total parking in the lot is from 501 to 1000, the minimum number of accessible spaces shall be 2% of the total. For a total number of spaces = 630, this results in 12.6 spaces. Provide 13 spaces.

Answer is D

Solution 438

Chapter 18 of the Highway Capacity Manual (2010) provides the following guidance for pedestrian green time:

$$t_{ps,do} = 3.2 + \frac{L_d}{S_p} + 0.27 N_{ped,do} \quad for\ crosswalk\ width\ W_d \leq 10\ ft$$

where $N_{ped,do}$ = number of pedestrians waiting at the corner to cross major street, which is given by

$$N_{ped,do} = N_{do} \frac{C - g_{walk,mi}}{C}$$

N_{do} = number of pedestrians arriving at the corner each cycle, which is given by

$$N_{do} = v_{do} \frac{C}{3600} = 400 \times \frac{75}{3600} = 8.33$$

$$N_{ped,do} = 8.33 \times \frac{75-15}{75} = 6.67$$

$$t_{ps,do} = 3.2 + \frac{48}{3.5} + 0.27 \times 6.67 = 18.7\ s$$

Answer is A

Solution 439

Answer is A (MUTCD section 6G.02)

Solution 440

At the rate of return (i), the present worth should be zero.
The $350k capital expenditure is a present value (P).   NEGATIVE
The $25k reduction in annual costs is an annuity (A)   POSITIVE
The $200k increase in salvage value is a future sum (F)   POSITIVE

Converting all of these to present worth, the net present worth can be written:

$$PW = -350 + 25 \left(\frac{P}{A}, i, 20\ yrs\right) + 200 \left(\frac{P}{F}, i, 20\ yrs\right) = 0$$

For i = 5%, PW = 37k
For i = 6%, PW = -0.9k   Actual answer 5.97%
For i = 7%, PW = -33.5k

Answer is B

Made in the USA
Middletown, DE
30 September 2019